YOUR KNOWLEDGE HAS VALUE

Carol Benjamin

Bias and Confounding in Research

GRIN Verlag

Bibliografische Information der Deutschen Nationalbibliothek:

Die Deutsche Bibliothek verzeichnet diese Publikation in der Deutschen National-
bibliografie; detaillierte bibliografische Daten sind im Internet über http://dnb.d-
nb.de/ abrufbar.

Imprint:

Copyright © 2010 GRIN Verlag GmbH
Druck und Bindung: Books on Demand GmbH, Norderstedt Germany
ISBN: 978-3-656-55126-3

This book at GRIN:

http://www.grin.com/en/e-book/265473/bias-and-confounding-in-research

GRIN - Your knowledge has value

Der GRIN Verlag publiziert seit 1998 wissenschaftliche Arbeiten von Studenten, Hochschullehrern und anderen Akademikern als eBook und gedrucktes Buch. Die Verlagswebsite www.grin.com ist die ideale Plattform zur Veröffentlichung von Hausarbeiten, Abschlussarbeiten, wissenschaftlichen Aufsätzen, Dissertationen und Fachbüchern.

Bias and Confounding in Research

Module 5 (SLP)

Carol Benjamin

TUI University

Abstract

This assignment involves identifying limitations of the proposed study. The potential bias

and limitation of the study is described. These involve proposed Study Design, proposed

Sampling Method, proposed Data Sources/Data Collection Methods, proposed Analytic

Methods. Finally the potential confounding factors in the study is discussed.

1. Limitations of proposed Study Design. Discuss potential bias and limitations.

The proposed study design that is selected for the study is a case control design. This design will be very effective in exploring the predictors of severe infection due to H1N1. However an article by Geneletti et al (2009) argues that retrospective case–control studies are more susceptible to selection bias than other epidemiologic studies as by design they require that both cases and controls are representative of the same population. However, as cases and control recruitment processes are often different, it is not always obvious that the necessary exchangeability conditions hold. Selection bias typically arises when the selection criteria are associated with the risk factor under investigation.

2. Limitations of proposed Sampling Method. Discuss potential bias and limitations.

In this study there will be some limitations and possibly bias of the propose sampling method. Since there will be limited data available for the H1N1 virus, the sample size will not be very large. There could be a selection bias since the charts that are reviewed will not represent the population. The two hundred charts that will be reviewed will not give a precise result. It is best to have a larger sample size and follow the cohorts over a longer period of time. Hackshaw (2008) reveals that the main problem with small studies is interpretation of results, in particular confidence intervals and P-values. Hackshaw also argues that when comparing characteristics between two or more groups of subjects (*e.g.* examining risk factors or treatments for disease), the size of the study depends on the magnitude of the expected effect size, which is usually quantified by a relative risk, odds ratio, absolute risk difference, hazard ratio, or difference between two means or medians. The smaller the true-effect size, the larger the study needs to be. This is because it is more difficult to distinguish between a real effect and random variation.

3. Limitations of proposed Data Sources/Data Collection Methods. Discuss potential bias and limitations.

The choice of data collection method which was proposed for this study will also have some limitation. The proposal stated that two hundred charts will be reviewed from patients who were admitted to the Emergency Room with a diagnosis of H1N1 Virus. There could be misclassification bias since some of the cases that are given an ICD-9 code for H1N1 might be incorrectly coded. Others cases that are true H1N1 cases could have received a different code. As a result of incorrect coding there could be differential or nondifferential misclassification. This could cause false association away from the null or mask an association when there is actually one.

4. Limitations of Proposed Analytic Methods. Discuss potential bias and limitations

The use of the non-parametric Mann–Whitney U test or chi-square test for linear trend will be used in the research. A P value of less than 0.05 is considered to indicate statistical significance. However these tests also have bias and limitation. The Mann-Whitney U test which is non-parametric is proposed to be used in this research to determine difference between groups. Nachar (2008) argues that there are limitations which can be seen when the sample size is similar or when the smallest manpower has the greatest variance, the t-test is more powerful on all the extent of the possible differences. Nachar also argues that the Mann- Whitney U test can give wrongful significant results. "The Chi Square Goodness of fit" article explains that the test can be applied for any distribution, However there are two limitations. The test is sensitive to how the binning of the data is performed and it also requires sufficient sample size so that the expected frequency is five.

5. Discuss potential confounding factors in your proposed study.

In studying predictors of severe H1N1 infection due to H1N1, there will be some limitation since the seasonal flu can be a confounding factor. Iannelli (2009) article reveals that the seasonal flu symptoms and H1N1 are usually the same and can include fever, cough, sore throat, running or stuffy nose, body ache, chills and fatigue. The author stated that the only difference is that children infected with H1N1 flu are more likely to have diarrhea and vomiting than those with seasonal flu. In the article "Influenza a H1N1" the author shows that the complications of H1N1 flu are likely to be similar to those of seasonal influenza which is most commonly pneumonia and respiratory failure. In the proposed study it could be difficult to determine the real causes of the predictors of H1N1. The seasonal flu could be the confounder since it can also contribute to pneumonia and respiratory failure.

Reference

Chi Square Goodness of Fit Test retrieved on June 14, 2010 from:

http://www.itl.nist.gov/div898/software/dataplot/refman1/auxillar/chsqgood.htm

Geneletti, S., Richardson, S., Best, N. (2008). Adjusting for selection bias in

retrospective case-control studies. Department of Epidemiology and Public

Health, Imperial College School of Medicine, London, UK.

Hackshaw, A. (2009). Small Studies: strengths and limitations retrieved on June 13,

2010 from: http://erj.ersjournals.com/cgi/content/full/32/5/1141

Iannelli, V. (2009). H1N1 Flu vs. Seasonal Flu retrieved on June 13, 2010 from:

http://pediatrics.about.com/od/swineflu/a/909_flu_facts.htm

Influenza A H1N1 retrieved June 14, 2010 from:

http://wiki.medpedia.com/Influenza_A_H1N1

Nachar, N. (2008). The Mann-Whitney U: A Test for Assessing Whether Two

Independent Samples Come From the Same Distribution. Tutorials in

Quantitative Methods for Psychology. 4(1) 13-20.